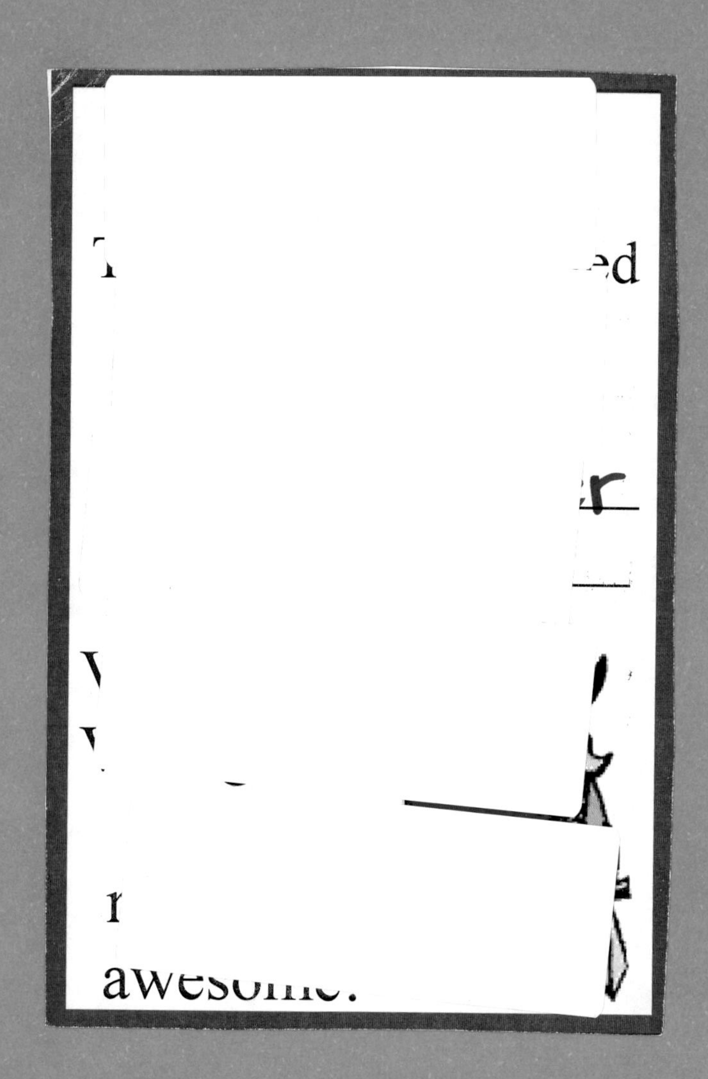

Jellyfish Inside Out

Michelle McKenzie

Monterey Bay Aquarium

The tentacles of a jellyfish provide shelter for these juvenile fish.

Contents

1. A jellyfish isn't a fish!

Have you ever found a jellyfish washed up on the beach? There's not much left to this animal when it's out of the water and has begun to dry up, is there? Just a jelly goo. That's because, unlike a fish, a jellyfish has no bony skeleton to cling to, no tough layer of scales to hold it together. Unlike a fish, a jellyfish has no heart, no brain, no blood.

lion's mane jelly

Crossota alba

A jellyfish is a simple animal made up of three main parts: some tentacles, a mouth, and a bell—that gooey, umbrella-shaped body. And it's 95% water! With only two thin layers of cells and some gel in between, both holding it all together, the jellyfish is pretty fragile.

At the Monterey Bay Aquarium, we group all gelatinous animals under the term of jelly. Comb jellies, gooseberry jellies, and Portuguese man-o-war look like jellyfishes. They have the same gel-gooey look, and some of them can sure sting you, but they differ from jellyfish in interesting ways. On the following pages are the four kinds of gelatinous animals we'll be talking about in this book. The diagrams will show you the differences between them.

sea gooseberry

sea nettle

Four kinds of gelatinous animals

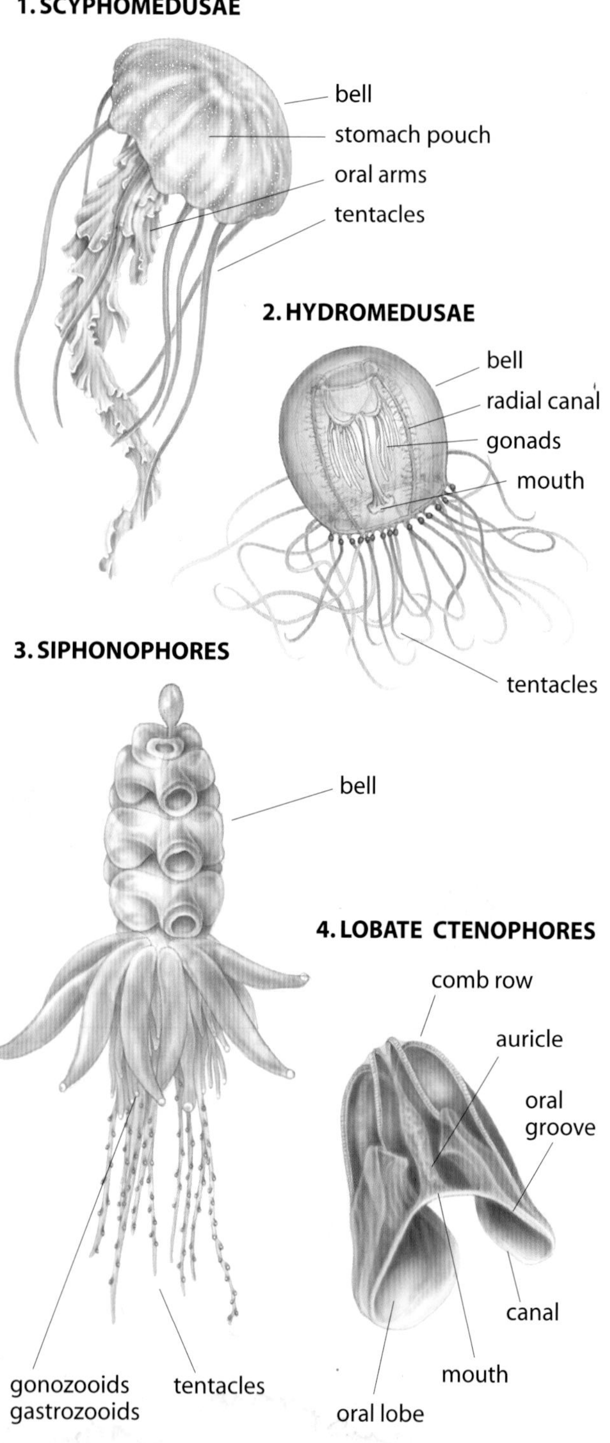

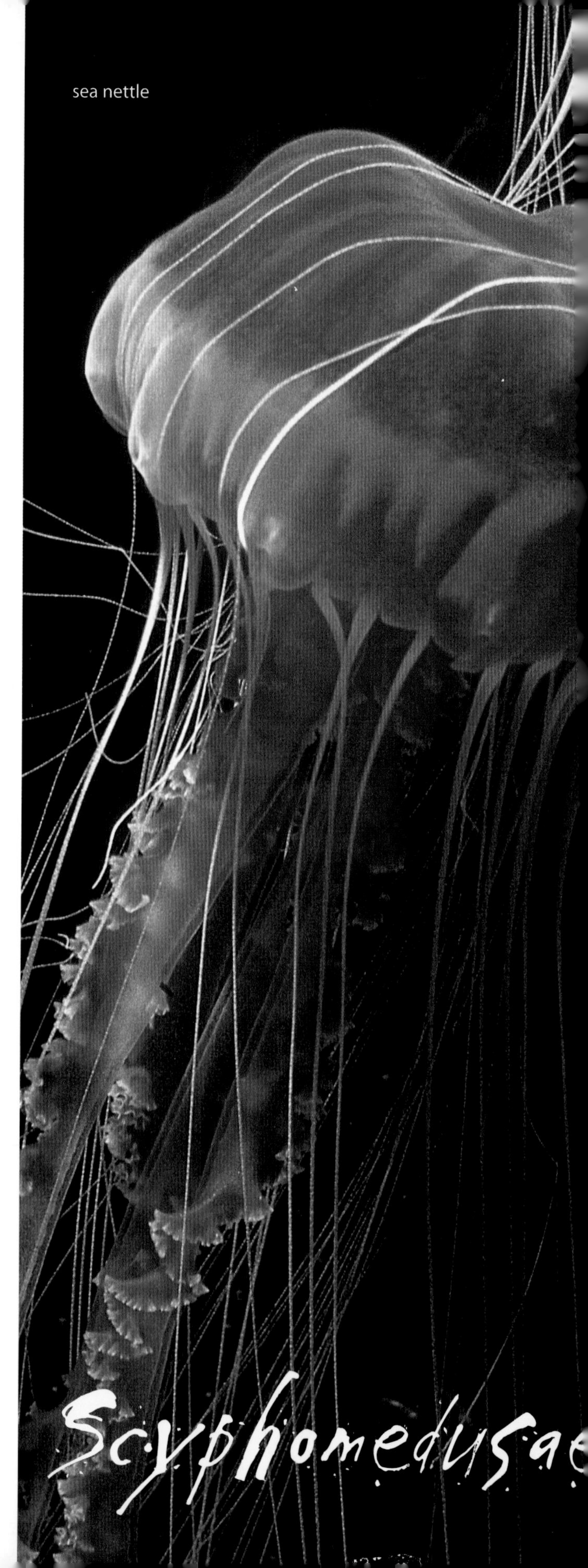

sea nettle

1. SCYPHOMEDUSAE, like the moon jellyfish and sea nettles, are considered the true jellyfishes. They're often large, with round bells and scalloped edges, and tentacles with stinging cells. Some have long, frilly oral arms near their central mouth.

2. HYDROMEDUSAE are a close relative to the true jellyfishes; they have a bell for propulsion and tentacles with stinging cells. Most of these jellies are small, the largest being about the size of a dinner plate. Compare that to the largest scyphomedusae, which can grow to the size of a dinner table!

3. SIPHONOPHORES are jellies that grow in long gelatinous chains. These are distant relatives to the jellyfishes. Some are very short and some are very, very long. Certain species, like the Portuguese man-o-war, *Physalia,* have potent stingers. Most prey on zooplankton and tiny fishes.

4. CTENOPHORES, (TEEN-o-fours), or comb jellies, belong to another phylum, and although they may look similar, they have some very interesting differences. They use hairs, called cilia, to paddle, oar-like, through the water. These eight rows of cilia diffract light into a colorful rainbow, which give the comb jellies a flying-saucer look. Another difference is that they use sticky cells, instead of stinging cells, to capture their prey.

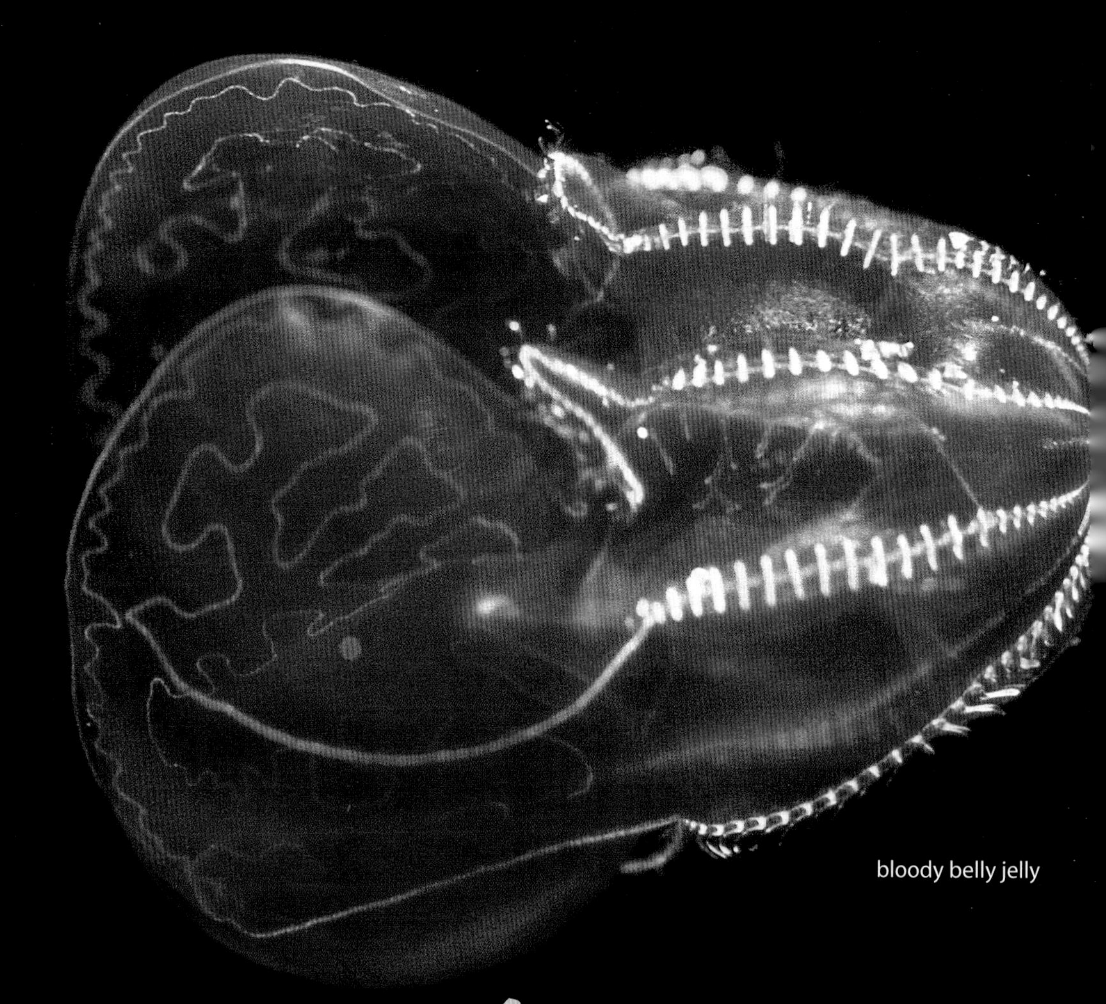

bloody belly jelly

Ctenophores

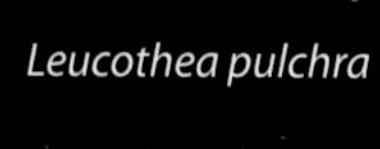

Leucothea pulchra

We will call them all jellies. And jellies fascinate scientists. There are 200 species of jellies in all sorts of colors, shapes and sizes. These mysterious creatures live in all the oceans and seas of the world, and even in some lakes. Some live near the surface of the water and others are found in the deepest oceans, traveling

200 species of jellies

alone or in groups. There are so many jellies that they make up the largest group of predators on Earth. A swarm of hungry jellies can gobble up huge amounts of small fishes and plankton.

spotted jellies

As we explore the lives of jellies, two scientists will share the latest jelly research with us.

Dr. Randy Kochevar and Dr. Judith Connor

Dr. Randy

DR. RANDY KOCHEVAR is a marine biologist who works at the Monterey Bay Aquarium in Monterey, California. Randy works with marine scientists from many different research institutions, and shares their research with aquarium staff members, reporters, writers, teachers and students around the world. Through the creation of web sites, CD-ROMs (including the one in this book!) and other outreach materials, Randy hopes to show people that the greatest hope for the oceans—and the humans that depend on them—comes through learning all we can about them.

Dr. Connor

DR. JUDITH CONNOR is a marine biologist at the Monterey Bay Aquarium Research Institute in Moss Landing, California. Using special equipment, like remotely operated vehicles (ROVs), video cameras and other cool tools, she observes and photographs deep sea animals never seen before. With the ROV, she can also collect these unusual jellies for study later in the lab.

The remotely operated vehicle (ROV), *Tiburon,* is used to explore the deep sea.

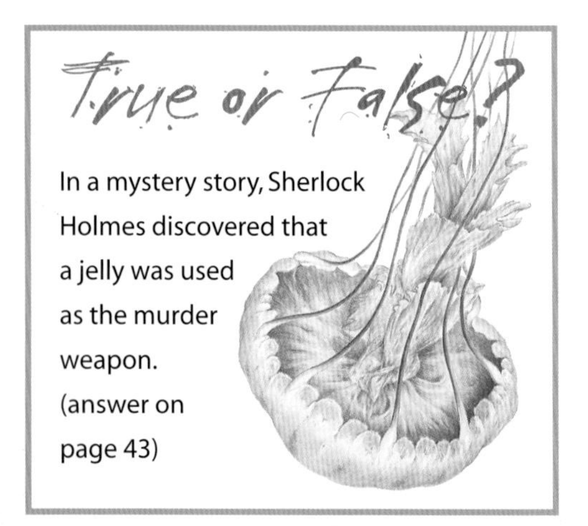

True or False?

In a mystery story, Sherlock Holmes discovered that a jelly was used as the murder weapon. (answer on page 43)

2. Caution!!

Imagine a harpoon that quickly pierces flesh with barbed hooks, then sends a hollow tube into the open wound to inject a poison so painful and toxic that it can kill. Sounds like a horror movie, doesn't it? But this is what most jellies do to eat and survive.

Imagine a painful, toxic poison

box jellies

Millions of stinging nematocysts

A jelly has thousands of harpoons, called nematocysts (na-mat'-o-sists). Its tentacles are covered with them! The good news is—once one of these poison-bearing harpoons has been fired, the jelly can't use it again. The bad news is—each jelly has a lot of nematocysts and they re-grow. The Portuguese man-o-war has around 2,000 of these stinging cells per inch. With dozens of tentacles, that adds up to millions of stingers!

Portuguese man-o-war

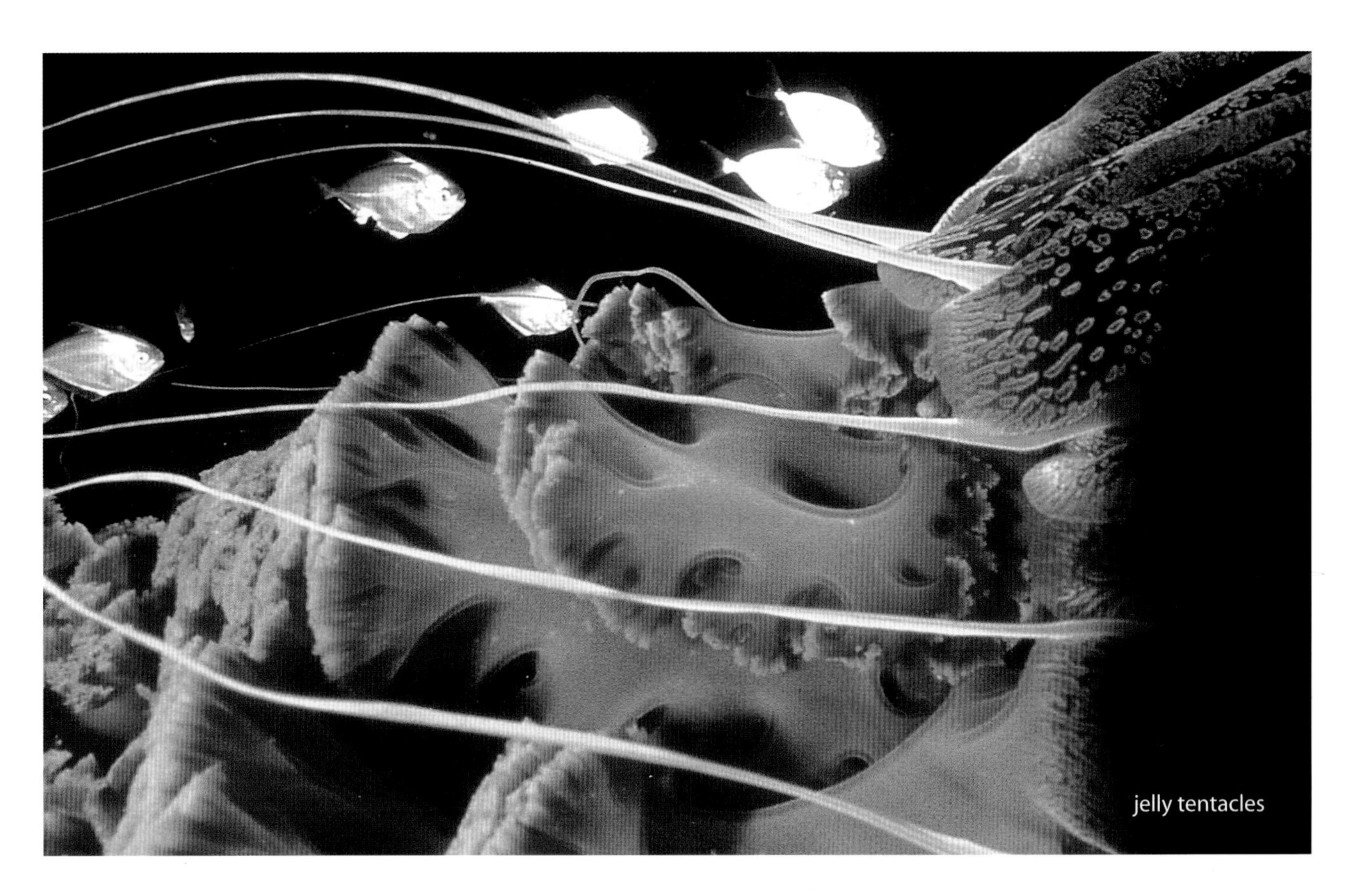

jelly tentacles

Some fish live directly under the jelly's bell, yet they avoid getting stung. Scientists originally thought these fish were immune to the stings, but they're not. They're just careful. If they venture into a tentacle, they too, would become a jelly's next meal. Jellies don't sting themselves or their own kind, though. They seem to sense when they've brushed up against one of their own tentacles or tentacles of their own species. Or even a glass wall. In these instances the stinging cells just don't fire.

Fish are very careful!

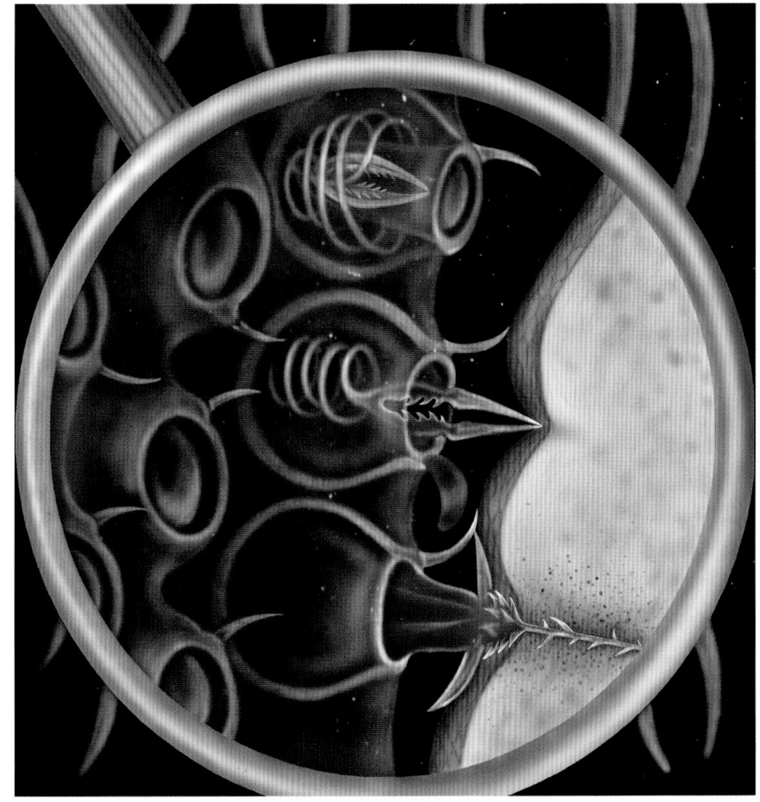

stinging cell action

Chrysaora fuscescens

Although a jelly's stinging mechanism is meant for capturing food and discouraging predators, thousands of people are stung each year, and for some, those encounters are deadly. Even when a jelly has been dead on the beach for awhile, the nematocysts are still dangerous. So when you see a jelly, it's best to admire its beauty from a distance.

Field Notes

Dr. Randy: "If you're ever stung by a jelly, use a cloth or gloves to remove the tentacles (they'll still sting you even though they're no longer attached to the jelly), and rinse the injury with sea water or vinegar. You'll have a pretty painful rash and maybe even welts where the stinging cells got you. Be sure to see your doctor if you have an allergic reaction, like shortness of breath."

Quiz

What's the most poisonous jelly in the world? (answer on page 43)

3. Jelly babies

moon jellies

Where do jelly babies come from? Different jelly species do different things to reproduce. Let's look at how true jellyfishes make more jellies. When the time is right, the male releases sperm and the female releases eggs.

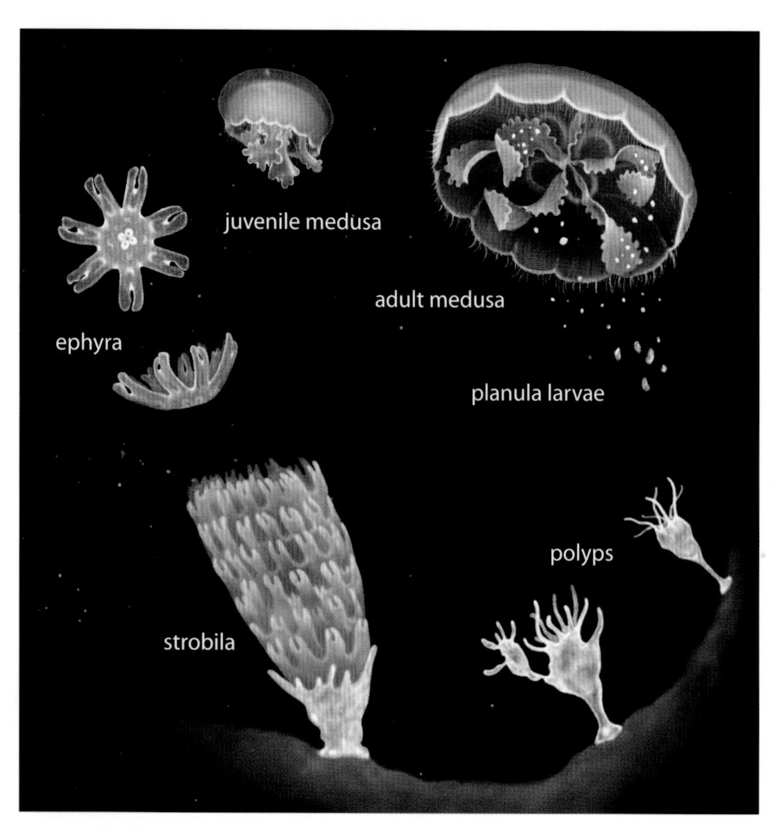

moon jelly life cycle

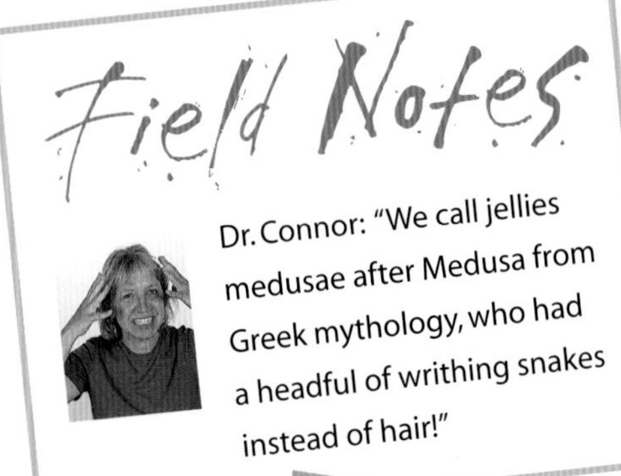

Field Notes

Dr. Connor: "We call jellies medusae after Medusa from Greek mythology, who had a headful of writhing snakes instead of hair!"

True or False?

Moon jellies went into orbit aboard the Columbia Space Shuttle in 1991 to see if jellies would eat french fries in a gravity-free environment. (answer on page 43)

The eggs and sperm mix together in the sea and once the eggs are fertilized, they form into tiny slipper-shaped creatures, called planulae, which swim to hard surfaces to settle down and grow.

Next, still on the rock, the planulae develop into polyps. The polyps begin growing tentacles that point upward. For a long time, no one knew that a polyp was actually a jelly because it looked so much like its cousin, the sea anemone.

A polyp can live for several years at this stage. It has a neat trick, too. It can clone itself by budding off more polyps, the way small branches grow on a tree. When the time is right, the next stage begins and the polyp grows a stack of many tiny jellies, one on top of the other, called strobila (and looks like a stack of pancakes). As the stack of jellies grows, a little jelly breaks free and swims off. Then the next one breaks free. And so on. In this way one egg can produce many jellies. Pretty amazing, isn't it?

Once the tiny jelly—or ephyra (e-FIE-ruh)—swims off on its own, it begins to grow mouth-arms and tentacles so it can drift and eat in the same manner as the adults. It's still pretty tiny at this point, maybe an 1/8 of an inch across, but as it grows, it looks more and more like an adult jelly.

planulae polyps and ephyrae

sea nettle polyps

egg-yolk jelly polyps

moon jelly ephyrae

egg-yolk jelly

Have you ever floated in the water as the waves and currents gently moved you out to sea or along the shore? You drifted away from your original position because currents move anything that isn't anchored. Similarly, most jellies drift along on these ocean currents, and pretty much go with the flow.

4. Smooth moves

pulse paddle propel

But is a jelly totally at the mercy of the currents? Actually, its bell helps it travel about in the water. Each pulse of the muscles in the bell forces water out and propels the jelly forward. Using this motion it can travel up toward the light, down into the darkest waters, and even along the surface, following the sun's movement across the water.

Jellies move in a variety of ways, too. Some, like the blue jelly, move quickly, with a pulsating rhythm. Others, like the sea nettle, move more slowly. Box jellies pulse their cube-shaped bells, flicking four clusters of tentacles as they go. A comb jelly paddles itself around by beating rows of "combs," or tiny hairs that act like oars, which move it through the water.

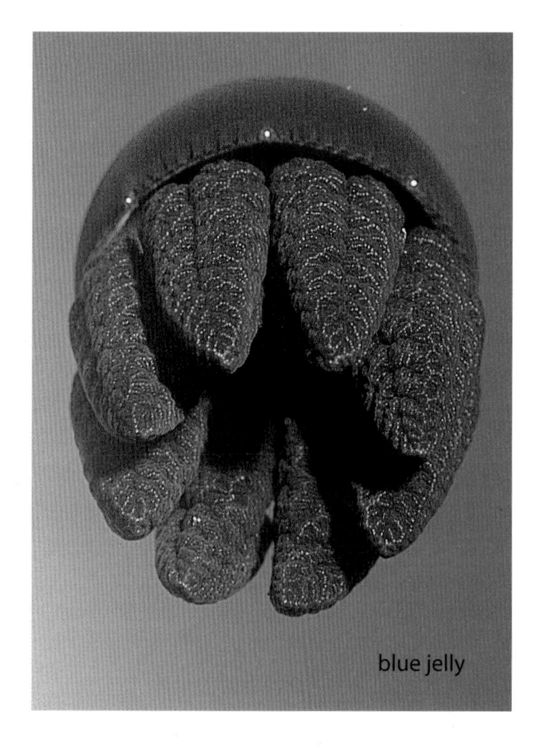

blue jelly

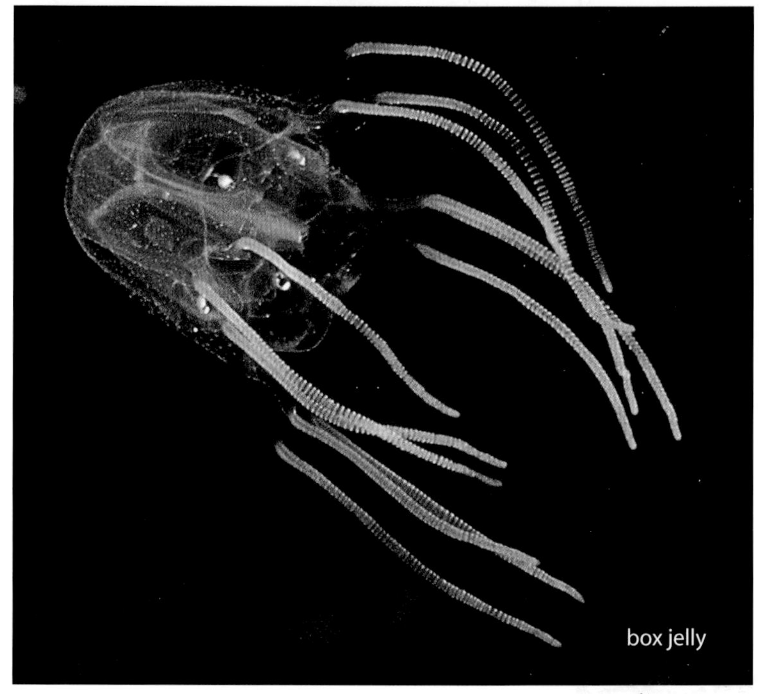

box jelly

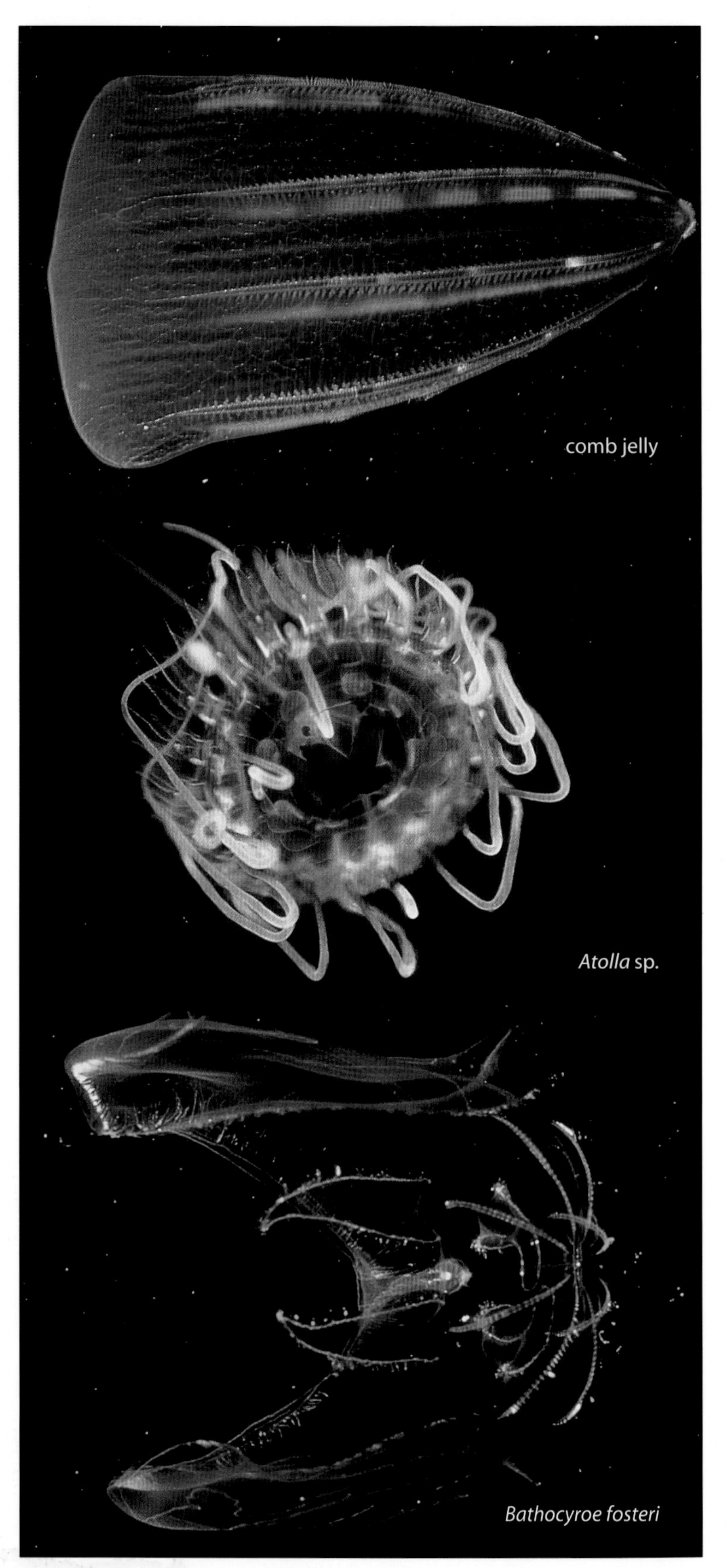

comb jelly

Atolla sp.

Bathocyroe fosteri

Field Notes

Dr. Randy: "Lots of folks believe that jellies have no brain. However, a jelly has nerves running throughout its body, and some scientists think this nerve net might actually be the jelly's brain—just spread out!"

Although it's not the brainiest creature in the sea, the jelly isn't totally without resources. Nerve cells signal the muscles in the bell that food or danger is nearby. The jelly can then move away from danger or toward the food source. Sensors around the rim of the bell let the jelly know if it's heading up or down, so it can correct its position as it needs to. Most jellies don't have real eyes either, just eye spots. While these eye spots don't form images, they might help jellies detect food and danger. When a jelly needs to travel toward the light, the eye spots help it find its way. The box jelly is an exception to this. It has amazingly complicated eyes that can actually form images!

True or False?

Some jellies give rides to tired sea creatures. (answer on page 43)

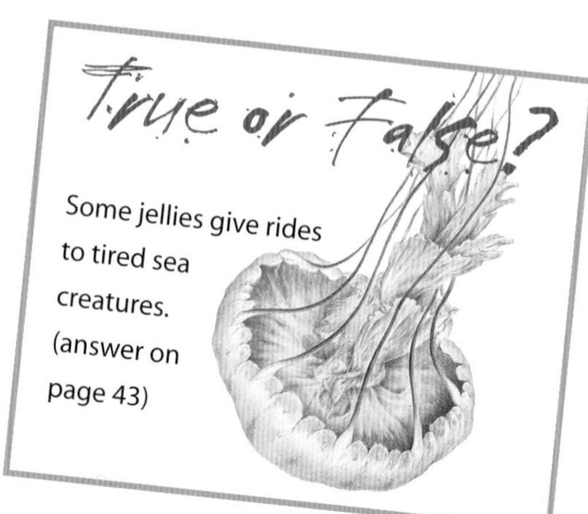

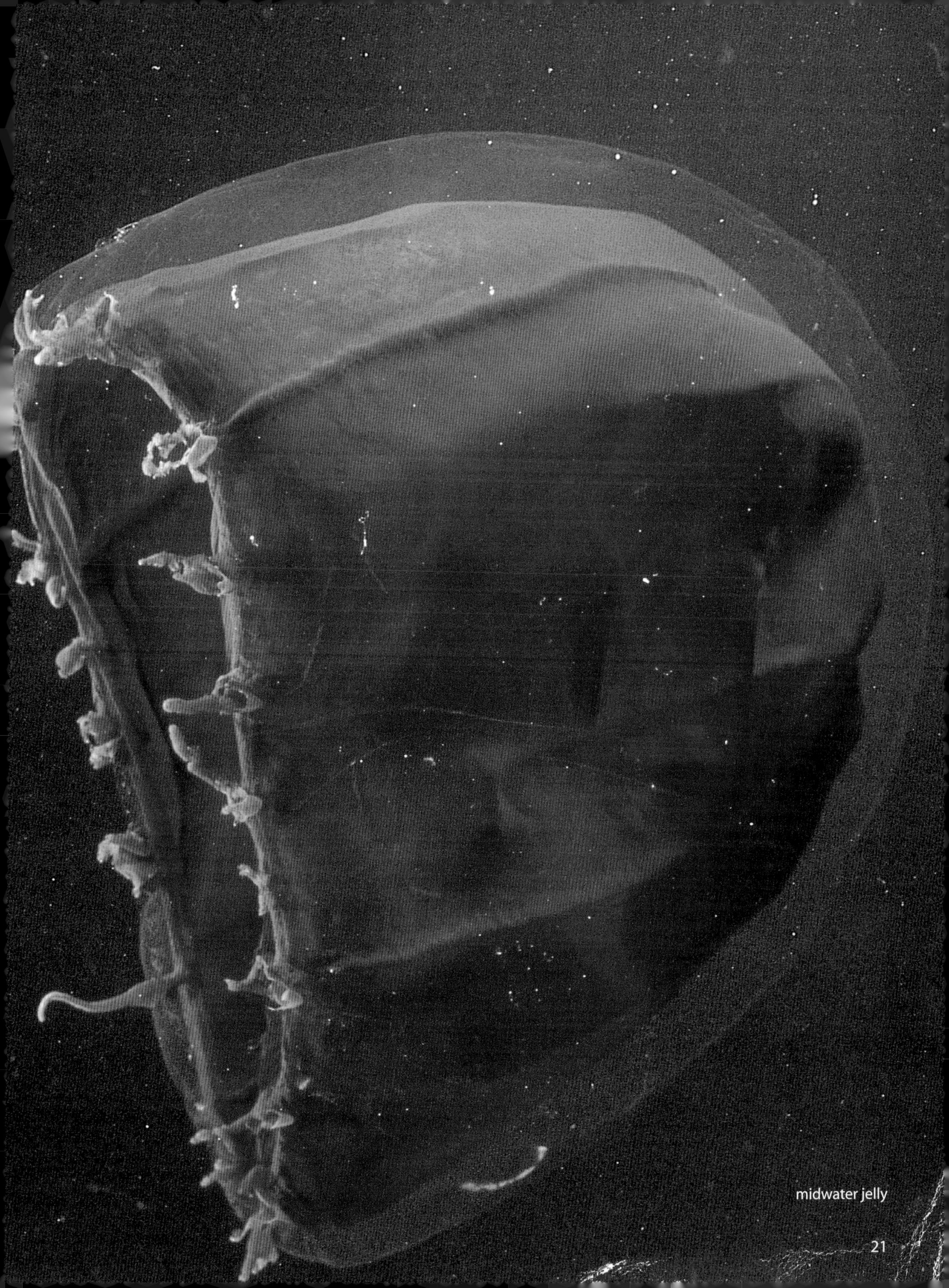
midwater jelly

5. Bulbous and huge or thin and stringy?

flower hat jelly

Jellies come in all shapes and sizes. You're probably most familiar with the classic umbrella-shaped jellyfish, such as the lion's mane, which is considered a giant by jelly

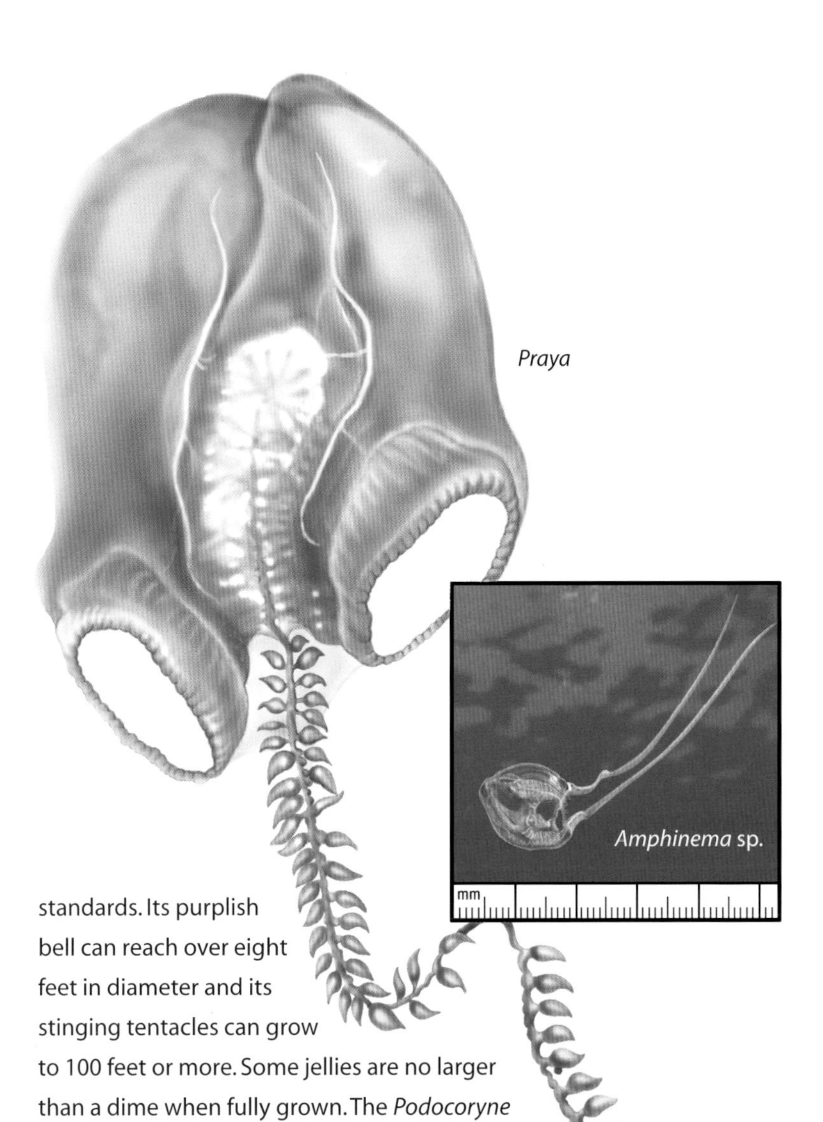

standards. Its purplish bell can reach over eight feet in diameter and its stinging tentacles can grow to 100 feet or more. Some jellies are no larger than a dime when fully grown. The *Podocoryne* sp. and the *Amphinema* sp. are only 2-3 millimeters in diameter, (about the size of a water drop, or less than half an inch).

Other gelatinous cousins look quite different from the classic jellyfish. The comb jelly has an oval-shaped body surrounded by rows of tiny, oarlike hairs, which it uses to paddle through the water. The siphonophore, *Praya,* is 120 feet long. That's even longer than a blue whale, and makes it the longest animal on Earth!

Jellies come in all shapes and sizes.

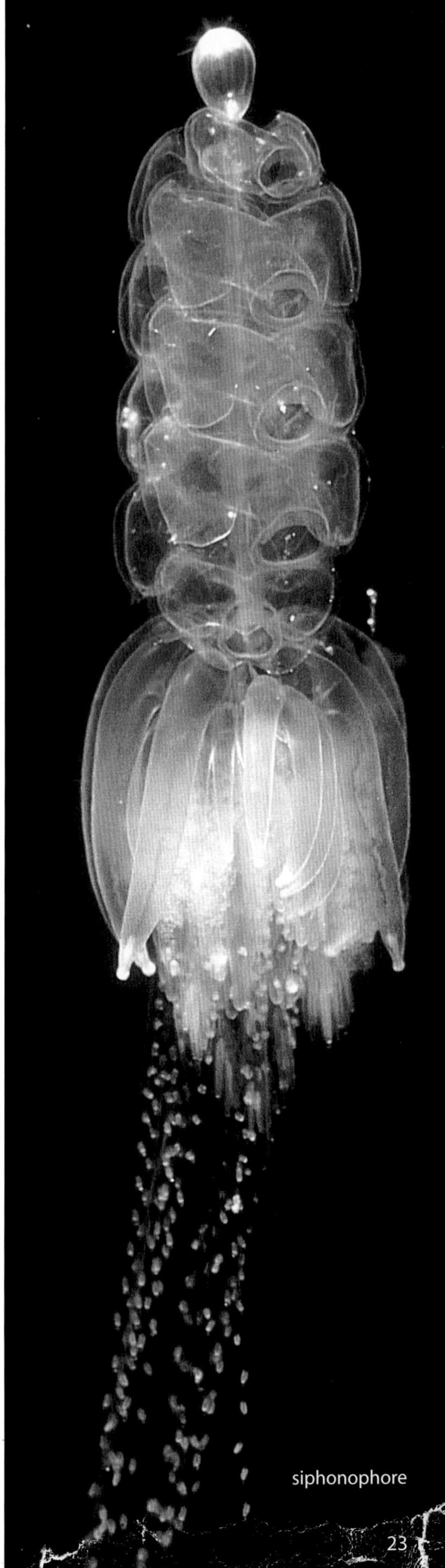
siphonophore

bloody belly comb jelly

Jellies catch prey in a variety of ways.

There are jellies with large, heavy tentacles for catching large prey. And other jellies with delicate, fringe-like tentacles for catching tiny fishes and plankton. Jellies that stay close to shore, to catch their favorite meal, are built to withstand the rough and tumble world of waves that drag them to the bottom of the sea. More fragile jellies stay in the calmer waters of lakes or estuaries, or out in the ocean's midwaters—far below the surface.

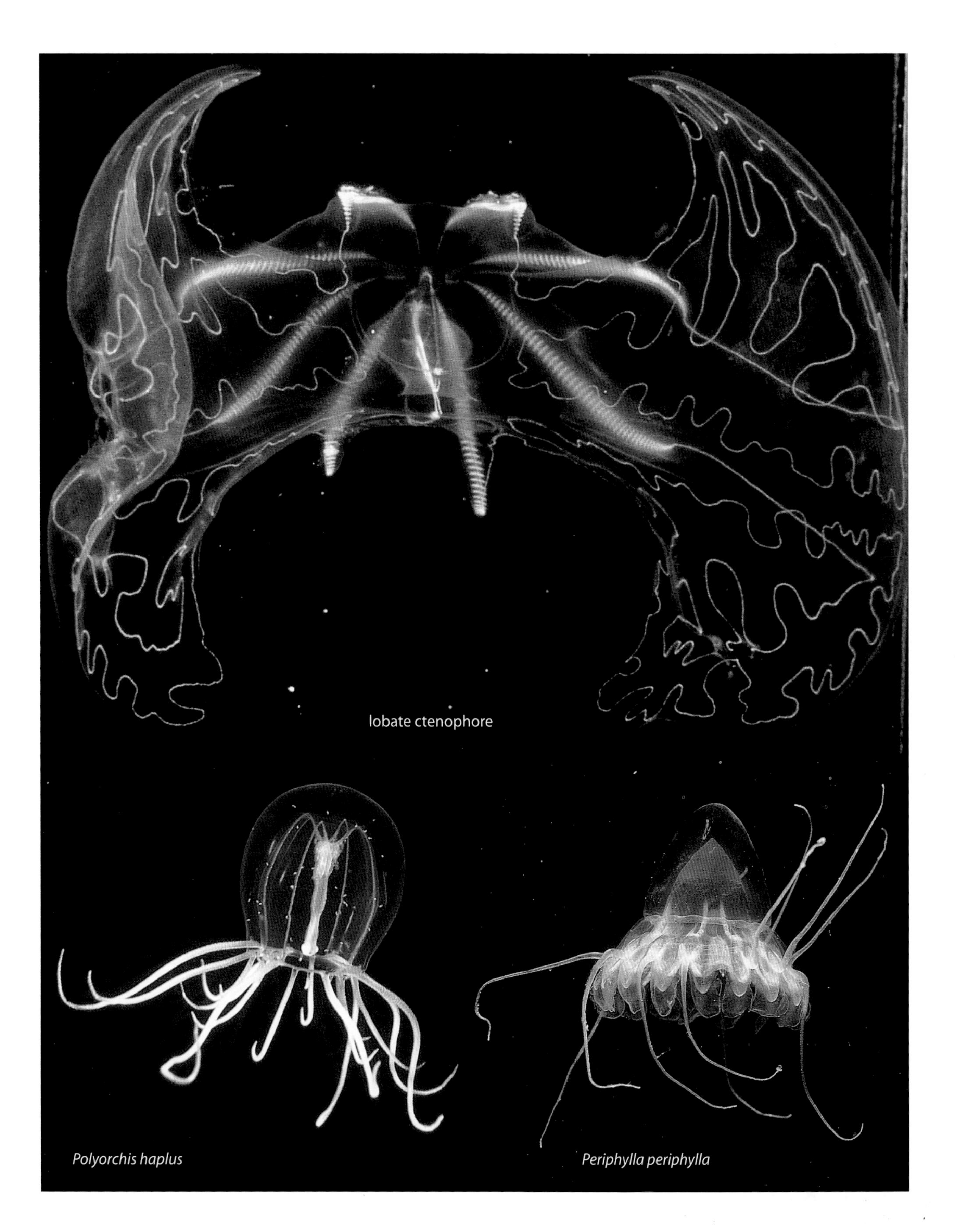

lobate ctenophore

Polyorchis haplus

Periphylla periphylla

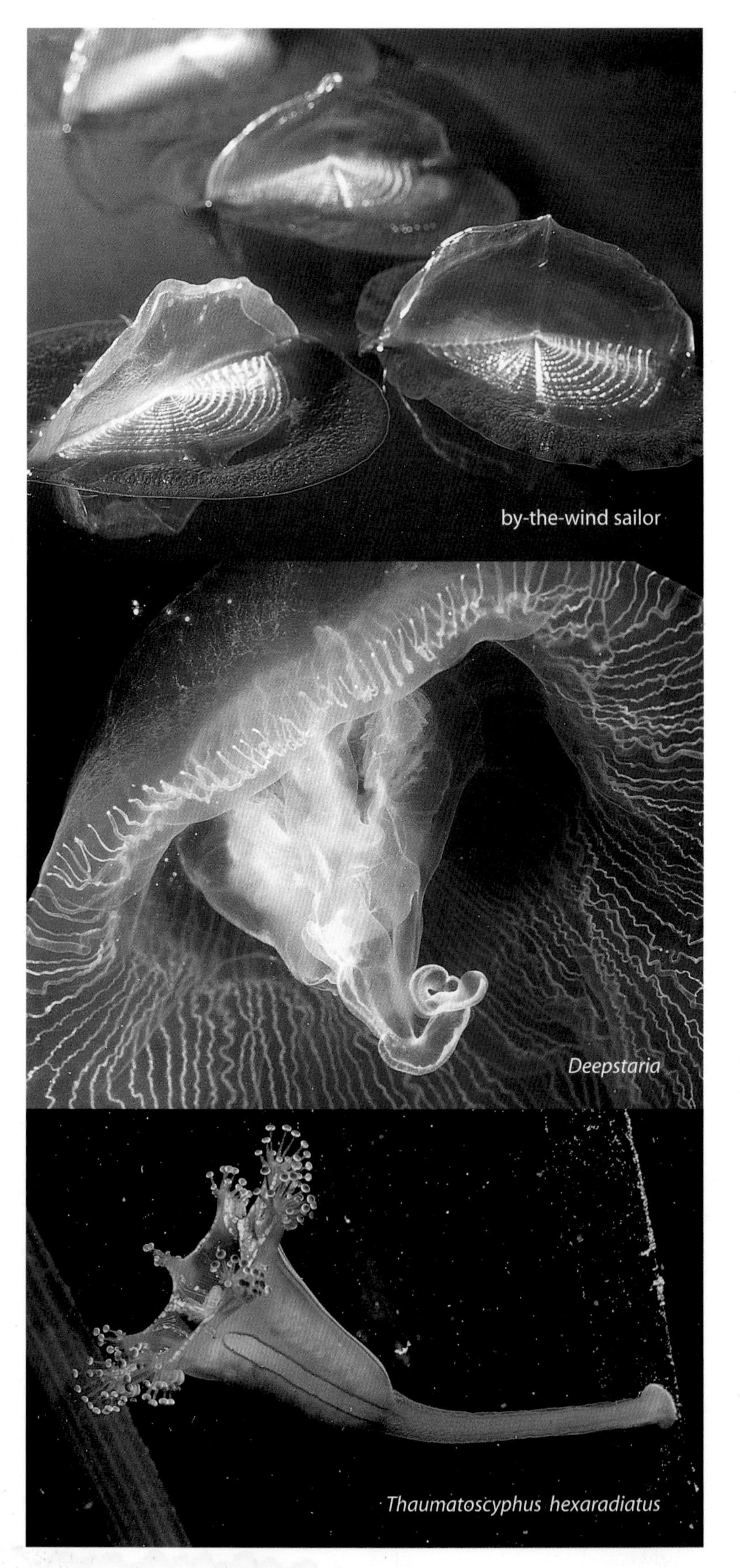
by-the-wind sailor

Deepstaria

Thaumatoscyphus hexaradiatus

Barrel shaped, umbrella shaped, long or short, tiny or huge, jellies manage to thrive in most types of environments. Jellies that surf on top of the water have floats for sailing along the surface, like the Portuguese man-o-war or the by-the-wind sailor. Others, like *Deepstaria,* stay in the deep, dark waters. All are well suited to the habitat that will support them and their food supply.

Field Notes

Dr. Connor: "How would you keep a record of the different jelly shapes you find? I like to draw and paint or photograph ocean critters that I see. I label each drawing with the date and place where the animal was found, then describe the habitat—whether it was growing on a rock or flung up on the sandy shore. Later, when I find something that looks familiar, I can look back through my drawings for similar shapes and colors I've seen in the past."

True or False?

A jelly will shrink when it gets wet. (answer on page 43)

6. Cool colors and pulsing patterns

spotted jelly

Green, pink, purple and red. Yellow dotted with cream-colored polka dots. There are many different jelly colors and patterns, and often there's a reason for this. As in the last chapter, it helps to think about where different jellies live, to figure out why there's all this variety.

pyrosome

crystal jelly

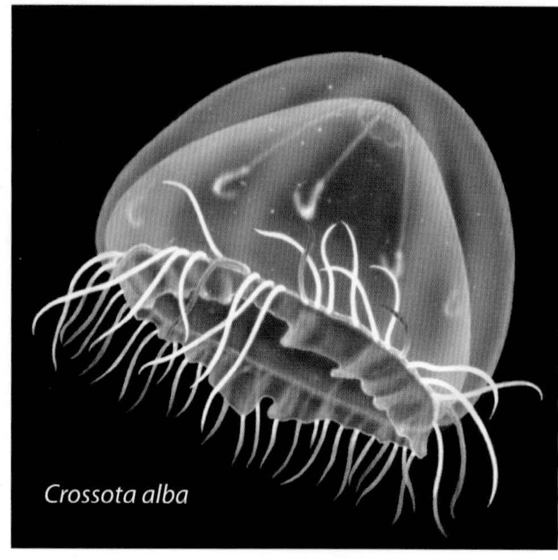

Crossota alba

Jellies that live up close to the water's surface or in the midwater are often nearly colorless and when viewed from below, they are probably invisible to potential predators. Deep sea jellies can be colorful, shades of purple and red. This, too, is clever camouflage because those bright colors don't show up in dark water.

Field Notes

Dr. Randy: "The chemicals that make jellies glow are being studied and used in laboratories and hospitals. It's helped doctors 'see' cancerous tumors during surgery. And one company is even putting these chemicals in squirt guns so they shoot glowing green water!"

Pleurobrachia bachei

Atolla vanhoeffeni

Ptychogastria polaris

Crossota alba

Some jellies make their own light!

Many deep sea animals, including some deep sea jellies, are bioluminescent. This means they can make their own light. Scientists know that this light comes from the reaction between a special chemical, called luciferin, and oxygen—but this reaction can only happen if an enzyme called luciferase is there, too. Some jellies may make luciferin themselves and others may get it from the animals they eat.

Scientists do wonder though, why clear jellies have dark-colored guts. They think those dark guts help hide the lights of glowing prey the jellies have eaten. A glowing stomach after a meal is a dead giveaway to a jelly's location. And jellies are on the menu for other creatures, such as sea turtles, ocean sunfish and even other jellies. Their sting doesn't seem to bother these predators, either. So being hard to spot is the best course of action if you're a jelly.

True or False?

The crystal jelly has glowing eyes. (answer on page 43)

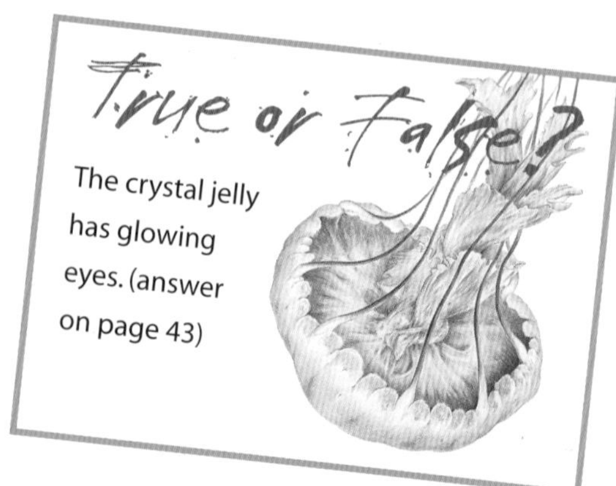

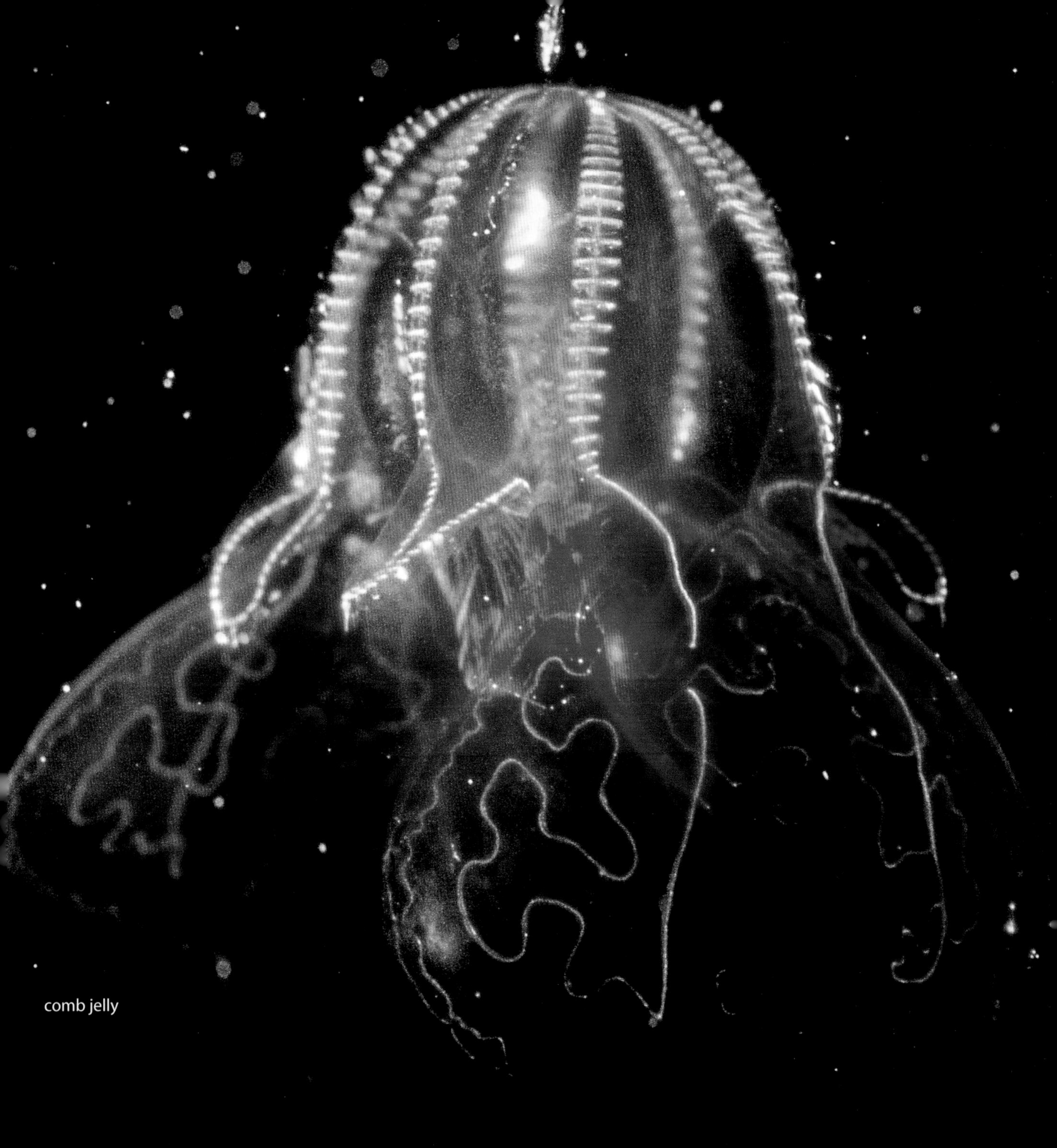

comb jelly

These eight rows of cilia diffract light into a colorful rainbow.

7. What's for lunch?

You may wonder how these creatures hunt. Some folks thought that jellies just snagged and ate prey that happened by—small fish, zooplankton, and even other jellies. But jellies can be picky eaters, and often capture creatures they won't eat.

Solmissus sp.

So how do they catch food? Remember, the nematocysts are powerful tools for catching prey. The nematocysts are triggered by the touch of a passing fish or plankton (drifting animals) and fire harpoons. Then toxins are pumped into the wound and the prey is paralyzed. Paralyzing its prey makes it easier for the fragile jelly to eat without getting injured. The tentacles then bring the captured animal up to the jelly's mouth.

Scientists wondered why some jellies swim at all—using up their energy—when they just seem to stay in the same spot. But recently it was discovered that the round shape of the pulsing bell creates currents in the water, called vortices. The bell pulses and the water flows under it and back out through the tentacles. This flow drags prey to the tentacles, where they get snagged for a meal. So while it doesn't get to choose what's there, the jelly's chances of catching something to eat are improved.

But while some jellies stay put, and others just drift, *Solmissus* spp. is an active hunter, by jelly

Carinaria cristata

Stomolophus meleagris

This pulsing movement of the bell creates vortices.

upside-down jelly

standards, and is almost always on the move. As it pulses through the water, it spreads its tentacles up and out to the sides. With its tentacles over its bell it's able to approach quickly moving prey and capture a meal.

The upside-down jelly, *Cassiopeia,* is a farmer. It lies upside down in shallow waters of tropical lagoons, pulsing against the seafloor. Algae grow inside the bell of this jelly. The jelly's natural inclination to swim upside down exposes the algae to the sunshine it needs to grow. In return, *Cassiopeia* gets to absorb the nutrients released from the algae. This symbiotic relationship has only one problem: if the algae die the jelly will begin to consume its own tissue and could die as well.

Field Notes

Dr. Randy: "Plankton is made up of tiny, drifting plants and animals. Plant plankton is called phytoplankton (FIE-toe-plankton), and animal plankton is called zooplankton (ZOH-ah-plankton). Because jellies are drifters, they fall into the zooplankton category and are tasty morsels for other animals."

Quiz

Do jellies have a sense of smell? (answer on page 43)

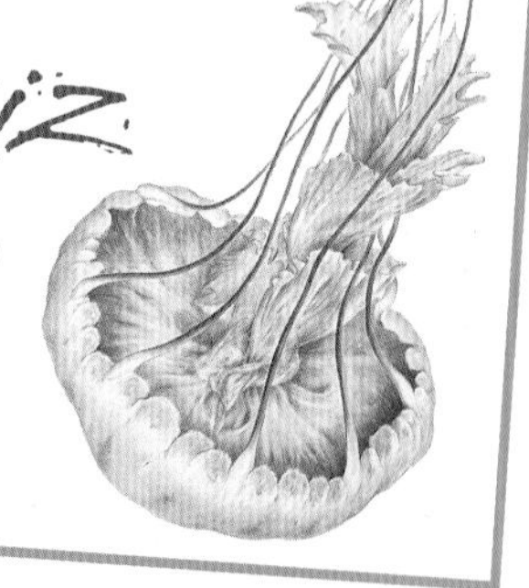

8. So you Want to be a Scientist?

Scientists as far back as the sixteenth century (that's in the 1500s) started a system of classification, called taxonomy, to help them keep track of all the amazing life on Earth.

Chrysaora melanaster

Feeding an octopus can be an eventful experience.

collecting jellies

They grouped animals and plants together based on similarities. For example, humans, birds, fishes, frogs and snakes all have backbones, so they are all grouped together as "chordates." Sea stars and sea urchins all have spiny skin and tube feet, so they are grouped together as "echinoderms," which means "spiny skin." The jellyfishes are in a group with anemones and corals, because all of these animals have a very simple body plan (where the food enters and leaves in the same place), and they all have special stinging cells. This group is called "cnidarians," (ny-DARE-ee-uns) which means "nettlelike."

All living things are organized into the following categories: **kingdom, phylum, class, order, family, genus, species.**

kingdom
phylum
class
order
family
genus
species

You can remember that by using this saying—"King Phillip came over for ginger snaps."

"King Phillip came over for ginger snaps."

An aquarist hand feeds jellies.

Here's how humans are classified:

kingdom - *Animalia*
phylum - *Chordata* subphylum - *Vertebrata*
class - *Mammalia*
order - *Primates*
family - *Hominidae*
genus - *Homo*
species - *sapiens*

Each animal is given a scientific name that has two words, the genus and species. These names are usually Greek or Latin, and usually describe something about the animal. Genus and species names are written in italics or underlined and the genus name is capitalized while the species name is not.

The Student Oceanography Club, at the Monterey Bay Aquarium, brings students ages 11-17 together to explore science, art and other marine topics.

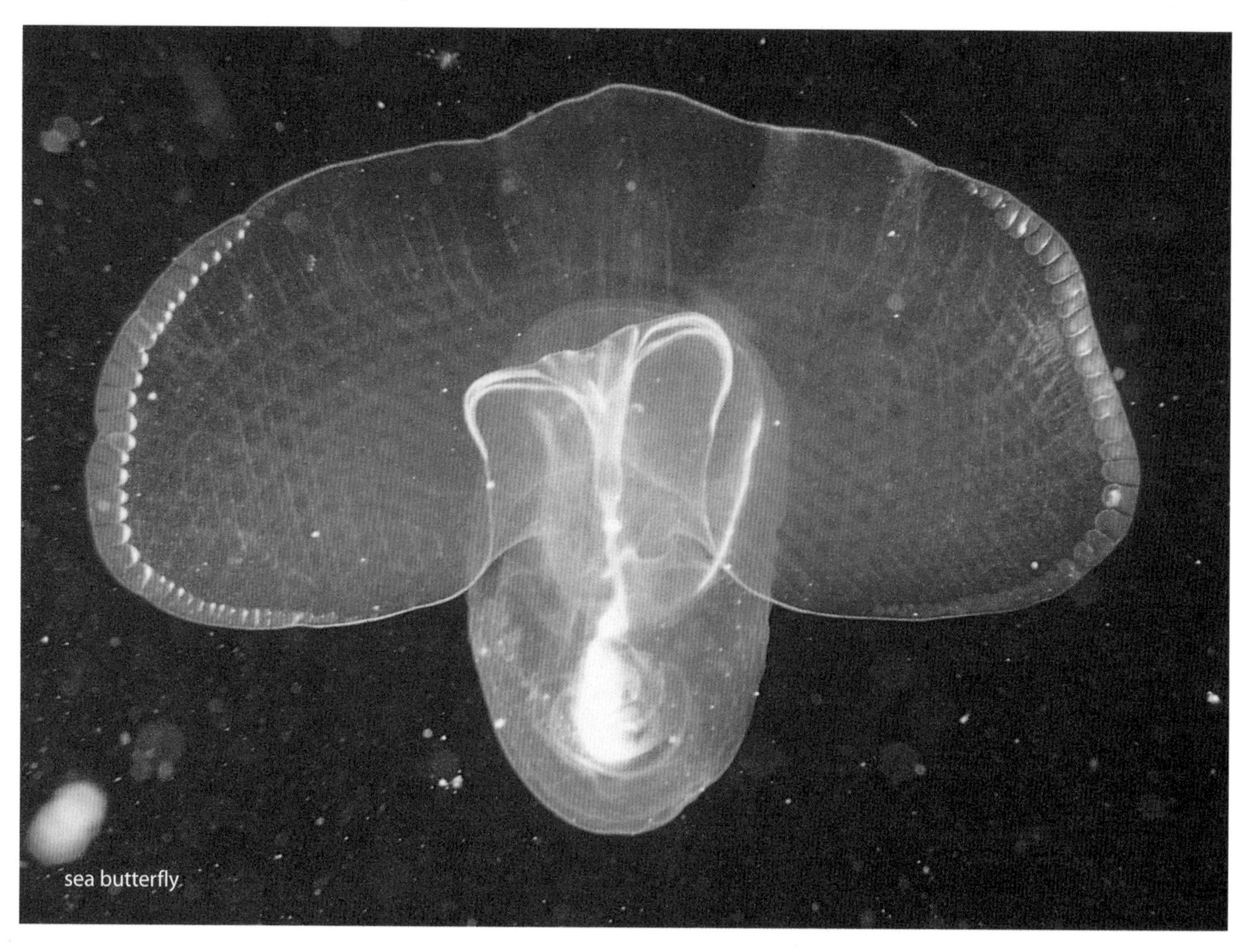
sea butterfly

Discover and name your own species.

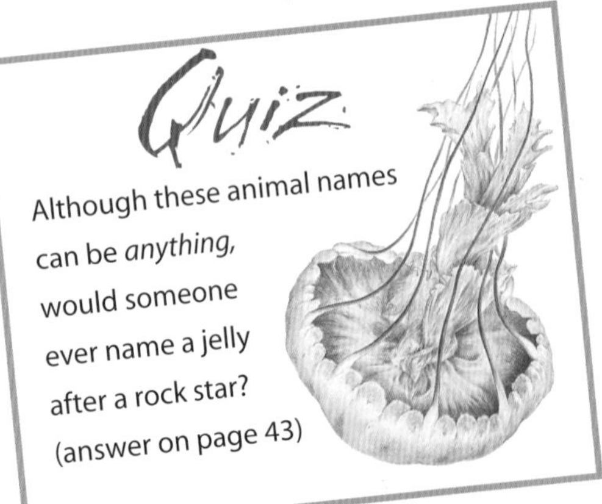

Quiz

Although these animal names can be *anything*, would someone ever name a jelly after a rock star?

(answer on page 43)

There's a reward in store for those who discover new species. The discoverers get to name the newly found species, and they can name them anything! Some are helpful reminders of what the animals look like, where they were found, or where they live. And some names are just plain silly and are named after mothers, dogs—even body parts!

There are also common names given to different animals. A common name is usually short and easy for people to remember. The scientific names are usually not short or easy to remember. For example, almost no one calls Chesapeake Bay sea nettles by their scientific name *Chrysaora quinquecirrha*. But not many jellies have a common name because people don't see them often enough to need to give them one.

Common names can be confusing too. The same animal may be known by a different common name in different parts of the world. Or the same common name can be used for different animals, such as the fishes called red snapper. There are 16 different fishes that are all called red snapper in different parts of the world. For example, the red snapper in the Gulf of Mexico is *Lutjanus campechanus,* but the red snapper of the US West Coast is *Sebastes* spp. (When talking about many different species within a single genus, use the genus name followed by "spp.," which means "multiple species.") These are two totally different types of fishes! You can see how using the scientific name helps you identify an animal, while the common name may lead to some confusion.

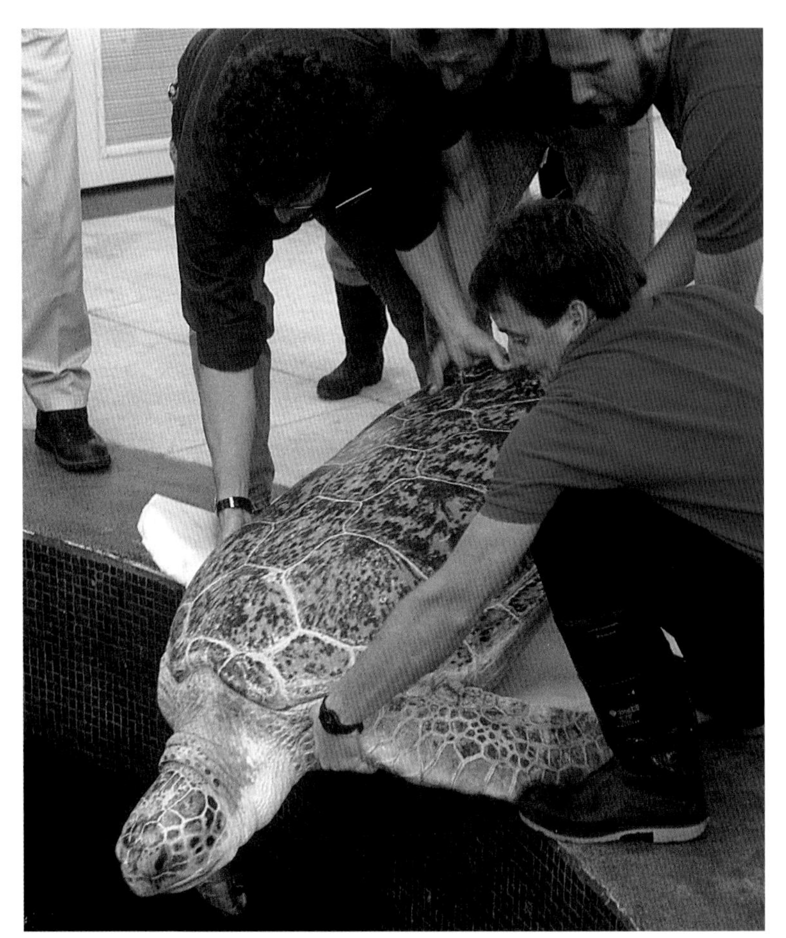

Aquarists release a sea turtle into an exhibit.

Dr. Randy: "Scientists' knowledge of deep sea jellies used to come from studying animals caught in nets. Jellies just break up into goo when netted, so it was pretty hard to learn much about them. Now, with the use of ROVs to help study life in the deep seas, we are beginning to understand how many different kinds of gelatinous animals there are. With ROVs to help explore the world of jellies, we're learning about their important role in the ecosystem."

An aquarist tows a plankton net through the water.

9. Ten ways to protect the oceans and the critters in it

blue jelly

The oceans are teeming with life, but many marine habitats and species are ailing because of human activities. You can help by following these ten steps.

1. **Be a picky eater.** Swordfish, shark and many other species are being caught faster than they can reproduce. And many fisheries accidentally catch sea turtles, dolphins, sharks and other animals along with the animals they mean to catch. Make sure the fish you eat are caught responsibly, and buy seafood recommended by the Monterey Bay Aquarium or other conservation organizations. To learn more, log onto www.montereybayaquarium.org

2. **Trim your waste.** Ordinary trash—such as discarded fishing line, plastic bags, and balloons—can choke or entangle marine animals. Pick up your trash and save a life!

3. **Contain that oil.** Did you know that more oil washes into the street drains and then into waterways through improper disposal of motor oil than from oil tanker spills? Oil can poison and kill plants and animals that depend on clean water. Make sure to quickly fix leaks, and find out where you can recycle used motor oil.

4. **Save energy.** Believe it or not, driving less and turning down the heat may help protect marine life. Rising global temperatures, which scientists attribute in part to fossil fuel burning, have had a devastating effect on many marine species in recent years. Higher sea temperatures have caused massive die-offs of coral reefs. So be part of something cool: save energy!

5. **Detoxify.** Toxic chemicals have seeped into every ocean on the planet, contaminating fishes, seals, polar bears and many other animals. Help reduce the toxic flow by choosing non-toxic cleansers, fertilizers and pesticides.

6. **Shop wisely.** Many marine products found in stores—from tropical fish to coral souvenirs—have been obtained illegally or have been harvested in ways that threaten marine biodiversity. Buy only those tropical products that have been certified as sustainably harvested by the Marine Aquarium Council. To learn more, visit www.aquariumcouncil.org

7. **Get smart.** Pay a visit to your local aquarium or curl up with a colorful nature book or watch a video about marine life. The more you learn, the better able you'll be to make smart decisions on behalf of the oceans.

8. **Spread the word.** Help other people get informed by writing to your elected officials, sending letters to your newspaper, and alerting your friends and family to the problems and wonders of the seas.

9. **Spare a dime.** Conservation organizations working on marine issues need all the help they can get. Send a contribution or join your favorite conservation group.

10. **Find joy.** Watch a sunset at the beach. Go on a whale-watching trip. Explore the wonders of a tide pool. Catch a wave. Sure, it's important to work hard to protect oceans. But remember to enjoy them too!

Be a picky eater
Trim your waste
Contain that oil
Save energy
Detoxify
Shop wisely
Get smart
Spread the word
Spare a dime
Find joy

Quiz

Balloons and plastic bags can end up in the ocean. Which endangered reptile is at risk of mistaking these for jellies and eating them? (answer on page 43)

doliolid

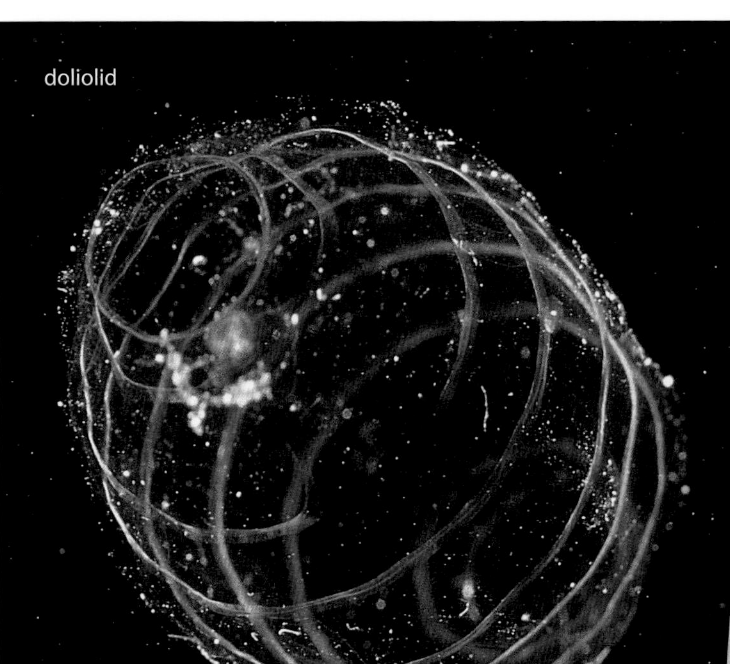

Field Notes

Dr. Connor keeps notes of her favorite habitats. "Pick your own special place and keep a record of creatures you see there. What do you see outside your window when you eat breakfast every day? A bird book from your local library will help you learn about the birds in your area. You can press leaves from the trees and bushes, or wildflowers from your yard or a nearby park. Count the number of each kind of tree. Record how many birds pass your window each week. Note when you see a weird bug or the first rain of the season. Your notebook becomes a witness to your changing world."

If you have questions you'd like to ask the scientists, you can e-mail Dr. Randy at dr.randy@mbayaq.org and Dr. Connor at dr.connor@mbayaq.org, or go to "web" on the CD-ROM and link to them from there.

After you've tried your hand at the crossword puzzle and search games, pop the CD-ROM into your computer and learn even more about jellies! You'll meet Dr. Randy and Dr. Connor on the "Meet the Scientists" page. Check out the glossary, download the screensaver, and while you're at it, give the match-up game a try!

A sea turtle snacks on a small jelly.

Answer Key

1. In a mystery story, Sherlock Holmes discovered that a jelly was used as the murder weapon. True. "Adventure of the Lion's Mane." Although the lion's mane sting can be fatal, most people survive their encounters with this gentle creature. (page 10)

2. What's the most poisonous jelly in the world? The Australian box jelly is the most dangerous jelly in the world. Its toxin is even more potent than a cobra and can kill a person in minutes! It also has some of the most well-developed eyes of any of the jellies. While most jellies merely sense light and dark, these guys have complex eyes with lens, corneas and retinas! (page 14)

3. Moon jellies went into orbit aboard the Columbia Space Shuttle in 1991 to see if jellies would eat french fries in a gravity-free environment. False. Actually moon jelly polyps and ephyrae went into orbit to help scientists observe the effects of weightlessness on internal organs. (French fries! Hope you didn't fall for THAT one!) (page 16)

4. Some jellies give rides to tired sea creatures. True! Well, sort of true. The egg-yolk jelly sometimes carries passengers, unintentionally, such as crabs and amphipods. And these passengers can damage the jelly. (page 20)

5. A jelly will shrink when it gets wet. False! How silly is that? But a jelly can shrink when there's less to eat, and then it can grow again as soon as there's more food. (page 26)

6. The crystal jelly has glowing eyes. Nope. It glows all around the margins of its bell, though. (page 30)

7. Do jellies have a sense of smell? Maybe. Scientists studying the cross jelly are finding that it might be able to smell food in the water. This means the cross jelly might actually pursue its prey because it smells it! (page 34).

8. Although these animal names can be anything, would someone ever name a jelly after a rock star? You bet. *Phialella zappai* was named after Frank Zappa by Ferdinando Boero, who named the new species he found, in hopes he'd get to meet the rock legend. And he did! (page 38)

9. Balloons and plastic bags can end up in the ocean. Which endangered reptile is at risk of mistaking these for jellies and eating them? Sea turtles eat jellies and plastic items, thinking they're jellies too. A meal like this can kill a sea turtle! (page 41)

Scrambled Jellies

Unscramble the letters to find a secret jelly word. The definitions are a hint.

HFSI	some jellies eat this	_ _ _ _
TGLHI	some jellies produce this to communicate	_ _ _ _ _
PPLYO	a baby jelly that grows after a larva settles	_ _ _ _ _
SLEPU	this body movement helps jellies get around	_ _ _ _ _
TRIFD	jellies do this to travel long distances	_ _ _ _ _
VRO	scientists use this to collect deep sea specimens	_ _ _
LRPUPE	being this color helps deep sea jellies hide	_ _ _ _ _ _
DDGUBNI	how a jelly polyp makes a clone of itself	_ _ _ _ _ _ _
GILAAEP	scientific name for purple jelly	_ _ _ _ _ _ _
YNCLOO	groups of members living together form this	_ _ _ _ _ _
LATEETCN	stinging cells are found along here	_ _ _ _ _ _ _ _
OHMTUAMR	another word for palp	_ _ _ _ _ _ _ _
ASYLPHAI	scientific name for Portuguese man-o-war	_ _ _ _ _ _ _ _
YASSEDWI	jellies swim up, down and this way	_ _ _ _ _ _ _ _

(answers on page 47)

Jellies Search

How many jelly words can you find? Be sure to look forward, backward, upside-down and diagonal, then circle the words you find.

C T A M E N D S E L T R U T P
S L S Y L L E J N O O M R A O
E S O O L C E L P R U P E N L
A S I N P H P D L L E J T N Y
M N T P E O S P I S C D A W T
R T A F H C E P O U A R W O S
E N S I P O A Y M T L I D D E
D O P R L M N L O U T F I E C
O I N E M A T O C Y S T M D I
N B O T H I J P P C O M E I T
I M V P O C E A N H O O M S R
H Y O Y L L E J B M O C T P O
C S R A V T E L I G A R F U V
E M A N O W A R F T L K E S I
C T N N O I T A V R E S N O C

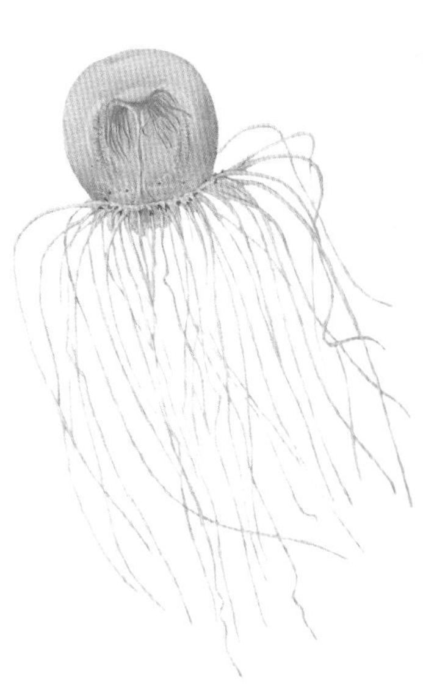

word list:

clone
nematocyst
drift
polyp
echinoderm
siphonophore
midwater
moon jelly
deep sea
upside down
fragile
ROV
man-of-war
turtles
comb jelly
ocean
fluid
purple
conservation
symbiont
vortices

(answers on page 47)

Crossed Up Jellies

Do you ever get your jelly facts crisscrossed? Here's a way to cross check your jelly knowledge.

Across

3. Predatory jellies catch and eat this.
7. Portuguese ___-of-war
8. Comb jellies don't have these so they can't see.
9. A clam has this hard body part for protection but a jelly doesn't.
11. This eight-armed swimmer is a jelly's neighbor in the sea.
13. A jelly ______ its body to swim short distances.
16. A jelly senses light with these.
18. This jelly sometimes has a live-in fish.
19. A jelly stores and digests prey here.
21. Is a squid a jelly?
22. Short for jellyfish
23. What you might say if you tasted and didn't like jellyfish salad.
24. Jellies produce this to communicate.

Down

1. This Australian jelly is the most dangerous to people.
2. A scientist's word for stinging cells.
3. Moon jellies are this rosy color.
4. Deep sea jellies live under a lot of this.
5. One use for stinging cells.
6. A jelly's umbrella-shaped body part.
10. This small jelly has rows of cilia but no stingers.
12. Jellies can tell the difference between __ and down.
14. A kid's word for stinging cell.
15. A siphonophore is a ______ of smaller members living together.
17. Most jellies live in the ___.
20. Palps pass prey to the ______.

(answers on page 47)

Answer Key to Puzzles

Scrambled Jellies

(page 44)

HFSI	FISH
TGLHI	LIGHT
PPLYO	POLYP
SLEPU	PULSE
TRIFD	DRIFT
VRO	ROV
LRPUPE	PURPLE
DDGUBNI	BUDDING
GILAAEP	PELAGIA
YNCLOO	COLONY
LATEETCN	TENTACLE
OHMTUAMR	MOUTH-ARM
ASYLPHAI	PHYSALIA
YASSEDWI	SIDEWAYS

Crossed Up Jellies

(page 46)

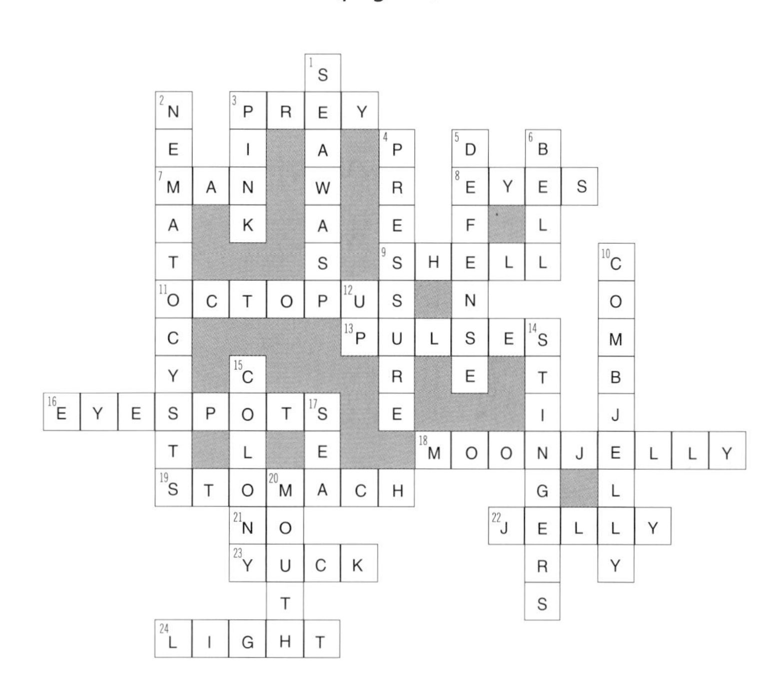

Alien Search

(page 45)

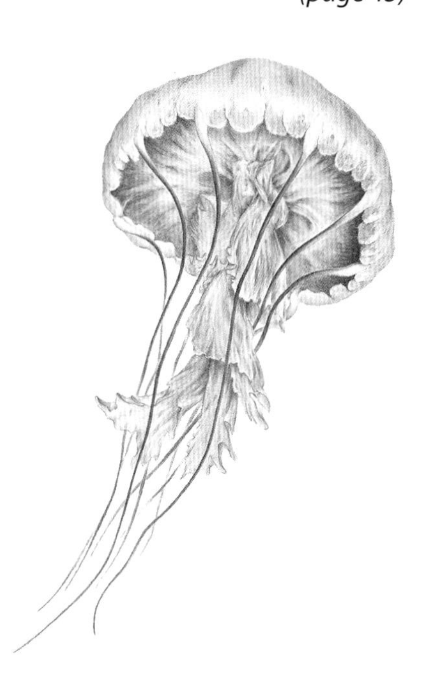

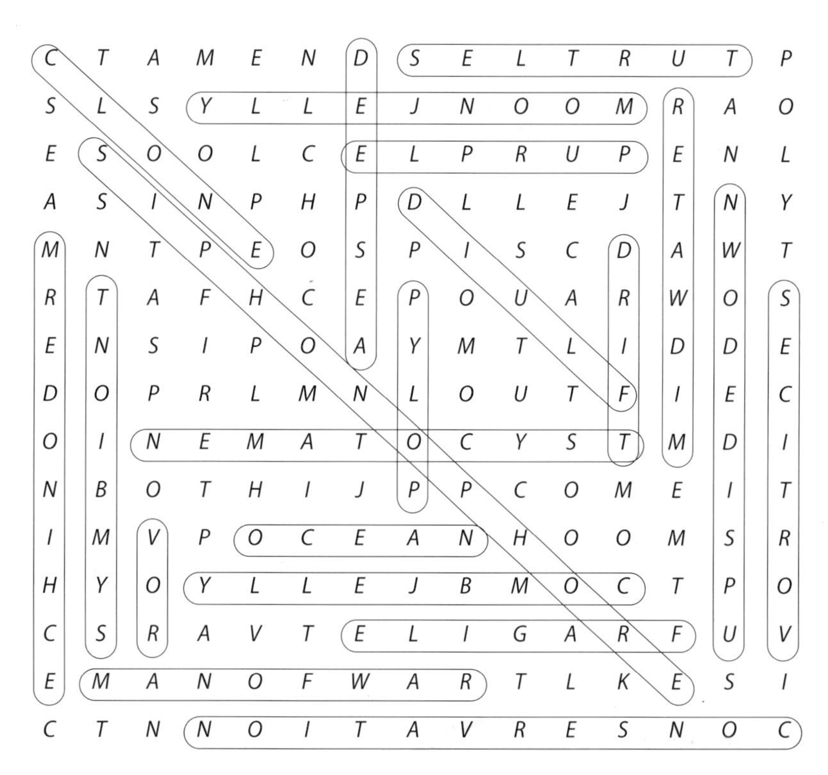

INDEX

SUGGESTED READING AND RESEARCH SOURCES

Websites:
Monterey Bay Aquarium www.montereybayaquarium.org
Monterey Bay Aquarium Research Institute www.mbari.org

Books:
A Guide to the World of the Jellyfish
Pacific Coast Pelagic Invertebrates
Jellies: Living Art
Shape of Life

Videos:
Jellies and Other Ocean Drifters
Life in the Deep

To find out more about related books and videos from Monterey Bay Aquarium Press visit our website at www.montereybayaquarium.org

Printed on recycled paper in Hong Kong by Global Interprint

DEDICATION

My warmest thanks to Judith L. Connor and Randy Kochevar for their support and contributions to this project. And a special thanks to Hank Armstrong, whose support made this project possible.

—Michelle McKenzie

Published in the United States by the Monterey Bay Aquarium Foundation, 886 Cannery Row, Monterey, California 93940-1085 www.montereybayaquarium.org

ISBN: 1-878244-43-4

Library of Congress Cataloging-in-Publication Date:
McKenzie, Michelle.
Jellyfish Inside out / Michelle McKenzie.
p. cm.
ISBN 1-878244-43-4
1. Jellyfishes - Juvenile literature. [1. Jellyfishes.] I. Title.

QL377.S4M35 2003
593.5'3--dc21 2002043121

MANAGING EDITOR: Michelle McKenzie
PROJECT MANAGER: Lisa M. Tooker
EDITORS: Lisa M. Tooker, Nora L. Deans, Miki Elizondo, Susan Blake
DESIGNER: Elizabeth Watson
WRITER: Michelle McKenzie
CONTRIBUTING WRITERS: Judith L. Connor, Randy Kochevar, Jane Cross
SCIENCE REVIEWERS: Steve Webster, Randy Kochevar, Judith Connor, George Matsumoto, Kevin Raskoff
ILLUSTRATOR: Marjorie Leggitt
JELLIES CD-ROM PROJECT MANAGER AND DEVELOPER: Randy Kochevar
JELLIES CD-ROM WRITERS: Ken Peterson, Karen Jeffries
JELLIES CD-ROM DESIGN: ad2.com

LITERARY CREDIT: 41, adapted with permission from the poster, *Joy to the Fishes and the Deep Blue Sea,* published by World Wildlife Fund, © 2002. All rights reserved. Visit www.worldwildlifefund.org for more information.

PHOTOGRAPHY CREDITS:
FRONT COVER: Steven Haddock;
BACK COVER: KEVIN RASKOFF (top Left); Steven Haddock (bottom right); Dave Wrobel/MBA (top right);
FRONT FLAP: Dave Wrobel;
BACK FLAP: Randy Wilder/MBA;
Bob Cranston/seapics.com: 13 (top); Ben Cropp: 34 (left), 42 (bottom); Steven Haddock: 30 (bottom); Howard Hall/howardhall.com: 9, 43; David Kearnes/seapics.com: 28 (top left); Jim Knowlton/seapics.com: 2-3; Larry Madin: 20 (bottom); George Matsumoto: 5 (top left), 20 (top left), 25 (top), 38; Claudia Mills/MBA: 28 (bottom left); MBA: 1, 10 (bottom right), 11, 24, 36, 37, 39 (bottom); MBARI: 32; Dawn Murray/MBARI: 20 (middle); Craig Racicot/MBA: 15; Kevin Raskoff/MBARI: 8 (top), 26 (middle left), 31; Charles Seaborn/MBA: 4, 26 (bottom left); Therisa Stack/Tom Stack & Assoc.: 12; Randy Wilder/MBA: 10 (top right), 14, 16 (middle), 20 (top right), 26 (right middle, 28 (bottom right), 34 (right), 39 (top); Dave Wrobel: 23 (right); Dave Wrobel/ MBA: 5 (bottom, right), 6 (right) 7, 8 (bottom), 17, 18, 19 (bottom), 21, 22, 25 (bottom), 26 (top left), 27, 29, 30 (top), 33 (right), 35, 40, 42 (top)